MW01178645

Father's Day
2000
Karen

PHYSICS IN THE REAL WORLD

Dr Jasper McKee

MINERVA PRESS
LONDON
MIAMI DELHI SYDNEY

PHYSICS IN THE REAL WORLD

ISBN 0 75410 816 3

First Published 1999 by
MINERVA PRESS
315–317 Regent Street
London W1R 7YB

Printed in Great Britain for Minerva Press

PHYSICS IN THE REAL WORLD

To my wise and wonderful grandchildren,
Christa and Colton

Acknowledgements

Thanks are due to Wanda Klassen for her accurate and insightful interpretation of a somewhat unruly text. Her attention to detail and her creativity have been invaluable in the production of this book. Gratitude is also due to Dr Hyman Gesser for his patience, fortitude, and efficiency in reviewing the final draft of the publication.

The author also acknowledges with thanks a bursary from the Royal Society of Canada which has assisted the publication of this book.

Foreword

This little book is a compendium of questions and answers relating to physics and physical science in everyday life. The questions are drawn from those submitted by listeners to local and national radio programmes sponsored by the Canadian Broadcasting Corporation. Whereas all the questions were solicited from listeners, the answers were provided on air by the local science correspondent, namely the author of this publication. The largely unscripted answers given on radio have now been reconstructed in prose form and are believed to be in essence those that were given at the time of each broadcast. Several additional questions offered on *Quirks & Quarks*, a weekly science programme on national radio, and questions from a special Christmas broadcast given in 1985 are included in the compilation. The intention of this book is to encourage the observational skills of readers and to indicate that there are many questions in the real world to which there are no perfect answers. The text is intended to be understood by most educated laypeople and students at high school and university. Many of the topics can be a matter of continuing discussion. Nonetheless, few follow-up questions were raised in response to solutions proffered on air, and it is believed that most answers contain the essence if not the whole truth of the solution to the problem. A glossary of useful words and their definitions is appended to the text of this publication.

From time to time doubts have arisen as to the veracity of the observations made by questioners. As a result, physicists at the University of Manitoba, and many friends of the author have carried out unusual tasks in order to verify the validity of a given question. Many quoted facts were counterintuitive in the first instance. Finally, thanks are due to the CBC for the opportunity to participate in an unusually interactive programme, and to Keran Sanders, Heather Reimer, Janet Dirks, Alison Hanks, among others, for their patience in hosting and producing the series.

Contents

Part One – Outdoors

Part Two – Theoretical

Part Three – Home and Kitchen

Part Four – Miscellaneous

Part One
Outdoors

Why is the environment so quiet after a snowfall?

Snow consists of crystals and flakes with a myriad of holes in between. To some extent it is similar to artificial forms of sound insulation or acoustic tiles. Sound is a pressure wave, and air molecules get lost and absorbed in the maze of little holes that is snow. The sound goes to heat up the surface of the snow, but with little success. Freshly fallen, highly porous snow is almost entirely surface and can have millions of square metres of surface per kilogram. Sound energy is absorbed on this surface.

While vacationing in England earlier this year, I was told that it would be a hot summer because the larks were flying so high. Is this merely an old person's tale, or is there a scientific basis for the forecast?

This turns out to be a straightforward question in physics. As the days get longer and spring arrives, air near the earth warms up and occasionally can become rather hot. In early summer when this occurs, the air higher up is still cold and the hot air rises more violently than usual. This air carries with it light insects, which are the food of the lark, and are forced upwards against gravity until the strength of the air current is completely spent. As a result, the lark's breakfast is laid out for him at a greater height than would normally be expected. The lark is therefore flying high not as a sign of coming hot weather but as an indication that air in contact with the earth is unusually warm. As a result the bird flies higher for an easy breakfast. The high flying lark is therefore not a harbinger of summer but rather the result of the warmth of air at ground level in late spring.

Unpowered barges floating down a river, the Rhine for example, often seem to move faster than the current. Indeed, the more heavily loaded the barge, the faster it goes. What is the physical explanation of this phenomenon?

There is a particular kind of friction that occurs only in gases and liquids. In such fluids, adjacent layers can move at different speeds giving rise to internal friction, drag or viscosity as one layer basically rubs against the next.

In a smooth-flowing river for example, the layer of water adjacent to either bank will essentially be at rest with respect to the bank. This will exert a drag on the next layer of water, which then slows the flow of the next, and so on. As a result, the centre of the river, assuming uniform flow and a uniform cross-section, will exhibit the most rapidly flowing water. On the other hand, whereas the bottom of the river creates drag in a similar manner to the bank, the surface is only in contact with air which is a gas and not a solid. Nonetheless, static air at the surface exerts some viscous drag on the water with the result that the velocity of the liquid is not at its maximum at the surface, but rather somewhere between surface and bottom, and indeed much nearer to the surface. For this reason, a heavily loaded barge, sitting deeply in the water will be moved along by a faster current. The lighter the barge, the higher will it float in the water, and the greater will be the effect of drag on the vessel as a whole. The surprising result is therefore easily understood in terms of viscous drag in fluids.

A gardener asks, which should be easier, pushing or pulling a wheelbarrow?

Pulling a wheelbarrow is easier.

Assume you are either pulling or pushing a wheelbarrow at an angle of 45° to the horizontal plane. In the case of pushing the wheelbarrow, the force exerted by the pusher at an angle of 45° to the wheelbarrow can be thought of as the sum of two separate forces, one pushing in the horizontal direction and the other acting vertically downward on the mass of the wheelbarrow. The effect of the downward component is effectively to increase the weight of the wheelbarrow and, therefore, increase the amount of friction experienced in moving the wheelbarrow forward. In the

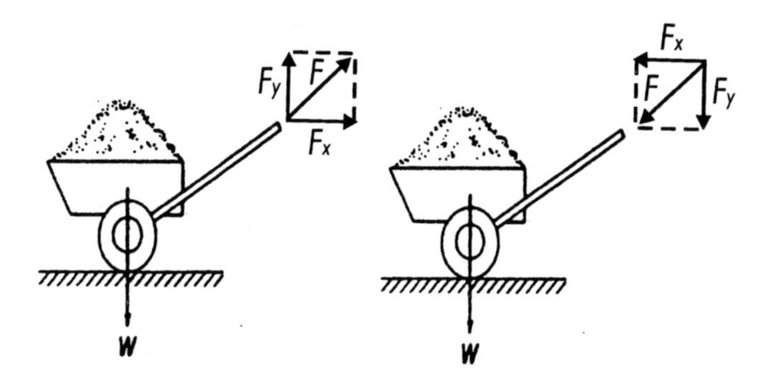

second case, in which the wheelbarrow is pulled, the force again acts at 45° and can be thought of as the addition of two separate forces, one again moving the wheelbarrow in the horizontal direction, which is what you want, and the other pulling on the mass of the wheelbarrow in a vertical direction which effectively reduces the weight of the wheelbarrow and thereby the friction between the wheelbarrow and the ground over which it travels. In short, pushing the wheelbarrow means that, as well as moving it forward, the wheelbarrow is pressed downward, effectively increasing its weight. In the case of pulling the wheelbarrow, the effect is the opposite. The force exerted tends to pull the wheelbarrow up vertically and this effect reduces the friction between the now lightened wheelbarrow and the ground. It is, of course, friction with the ground that always inhibits movement of the wheelbarrow whether forward or backward.

It is a fine summer afternoon, I am sitting relaxing in a chair in the yard and impressed by the outstanding beauty of the geraniums and roses in the flowerbeds that surround me. I fall asleep and only awaken as dusk is about to fall. At this time, I am overcome by the greenness of the grass and the leaves on the trees. What is the physical explanation for this dramatic change in my interpretation of the environment?

The answer to this question does not involve any change in the colour distribution of sunlight reaching earth. It is a matter of perception.

The human body contains two organs, eyes, which interpret,

among other things, the colour in our environment. Visible light coming to the eye, directly or indirectly from the sun contains a variety of colours from red, which is the least energetic, to violet, which is the most. The range of colours from red through orange, yellow, green, blue, and violet is known as the spectrum of visible light and the eye, as a detector, is not equally sensitive to all the hues or colours in the spectrum. Indeed, it perceives red and violet with great difficulty whereas yellowish orange is the colour most readily seen by the eye under most conditions of illumination. Infra-red or far red is invisible to the eye as is ultraviolet or far violet at the other end of the visible distribution. So, in general, the response of the eye, as it is called, to coloured light varies from poor at the ends of the spectrum, that's to say at both red and violet, to a maximum in the centre (normally in the yellow-orange region). Occasionally this peak moves however. Under conditions of low illumination where the intensity of light coming to the eye and stimulating the light receptors is low, the response (maximum sensitivity) of the eye moves from yellowish orange to yellowish green, making red even less detectable than in high illumination although somewhat enhancing the ability to see blue. It is this change in the overall response of the eye to colour that makes the red geraniums virtually invisible at dusk while the green of the grass is of outstanding visibility.

The receptors in the eye are known as cones of which there are three kinds and it is the interplay of these light receptors and their response to the intensities of different colours of illuminating light that determines the physiological response in the brain to the light falling on the eye. This phenomenon of the change in response of the eye with decreasing intensity of illumination was originally known as the Purkinge Effect after the researcher who first commented on the phenomenon.

When you look at an illuminated street sign, the red letters appear nearer to you than blue or green ones. Why is this?

The focal length of a lens depends on the wavelength of light. The lens of the eye focuses red light farther from the lens than the other colours. Indeed the red is focused slightly behind the retina giving rise to a slightly larger image at the retina. This gives the impression of the red letters being nearer to you than blue or green.

My rabbit, Honey, has a little plastic dish in which her water is left overnight. In late fall, I have noticed that if left overnight the dish becomes, not only frozen over, but frozen underneath as well so that a bubble of ice is left in the dish with liquid water in the middle. This seems a surprising thing to occur and I wondered what the explanation might be.

Once the outdoor temperature reaches freezing point it is likely that ice will form on even the purest water. As a result of air flow and surface evaporation the surface of the liquid can cool rapidly, causing ice to form and the density of unfrozen water to increase. Water is at its densest at 4°C, so as it cools it sinks. In the case of a plastic dish in close contact with cold ground, heat is lost quite rapidly from all areas of the dish in contact with the cold external environment. Water therefore freezes quickly both at the surface of the liquid, and where it is in direct contact with cold surroundings. The last of the water to freeze will therefore be near the bottom of the bowl where it will maintain a temperature of around 4°C until just before final freezing occurs. In the question, Honey's dish has frozen all round, but still contains some liquid water, which is what might be expected just before freezing is complete.

How is artificial snow created?

A tiny micro-organism called 'pseudomonas syringae' is added to small water droplets blown from a snowmaking machine. The water crystallises into snow more effectively, and at higher temperatures than normal – perhaps 2 to 5°C. As a result, snow is generated far more efficiently than otherwise.

The 'little bug' pseudomonas syringae is known to cause frost on plants. As water requires a particle like an ice crystal to crystallise upon in order to change into snow, this 'little bug' turns out to be particularly appropriate for the function. 'Lipoproteins' in the outer cell membrane of the bacteria are in a structure that looks like an ice crystal to the scientist and presumably also to the droplet of water.

A graduate student at Rochester University is believed to have been the first person to note the resemblance between an ice crystal and the bacterial structure. However, Advanced Genetic Sciences, Berkeley, California was the first industry to take advantage of this fact.

Why is aircraft noise so noticeable on clear cold nights? Does it also travel further on a cold than on a warm day?

Has aircraft noise been more noticeable lately? Yes. On most normal days in spring and summer air temperature decreases with elevation. Air is heated by the ground – not directly by the sun. However, on a calm, cold, clear night we can have a temperature inversion and temperature increases with height up to a certain elevation.

At night the ground radiates its heat to the air and no longer receives heat from the sun. If there are no clouds, the heat goes to higher elevations and stays there. As a result sound waves travelling from a low-flying aircraft are reflected downwards when the waves meet warmer air above their source. Because sound travels faster in warmer air, a sound beam moving at an angle away from the ground will find its top part moving faster as it meets warm air. The top moves faster and as a result the beam bends downward to earth where you hear it. It is like paddling a canoe – pull on the right-hand oar and you turn left – or take a cotton spool and let it roll from a smooth surface on to a rough one – at the boundary the spool will change its direction and move away from the smooth surface.

In contrast, on a warm day temperature decreases with height and no sound will be bent down to reach you. However climbers on a mountain can hear sounds from thousands of feet below due to the sound being bent upwards in that case. On 2 February, 1901, when canons were fired in London on the death of Queen Victoria, the sound was heard in London but not in the surrounding countryside. However the roar was heard distinctly in a village one hundred and fifty kilometres away – because of a temperature inversion and the bending downwards of the sound waves. The distance sound travels in total depends on various atmospheric factors. It is a pressure wave and temperature and air density are both important.

Avid tennis watchers must have noticed that many tennis players blow on their hands between shots. Is this a nervous affliction or do they actually cool their hands this way? If so, how can blowing on your hands cool them when the same action in winter seems to warm them?

The answer is that both phenomena are correct. Hands are warmed by blowing gently on them, and cooled by blowing hard from a short distance away! It is as simple as that. The normal temperature of air from human lungs is ~37°C. The temperature of the skin on the hands is however between 25 and 30°C. Blowing gently brings warm air from your lungs into contact with cooler skin and produces a feeling of warmth whether in winter or summer, if you do it correctly.

In contrary fashion, if you move your hand away from your mouth, ambient room air, normally at around 20°C, has more chance of being sucked into the air stream. This occurs because of Bernoulli's principle according to which the pressure in a moving stream of fluid (air) is less than that of the surrounding fluid (air). Low pressure sucks air in – as with a vacuum cleaner or indeed a shower curtain. The harder you blow, the more the cooler air is sucked into the breath stream, so the air reaching your hand is well mixed with outside air.

However, it is also significant that the fast jet of blown air breaks up the stagnant layer of air above the skin, which because of normal perspiration is saturated with water vapour from the body. Once the existing vapour is removed, further evaporation from the hand occurs, taking away heat and adding to the sensation of cold. The harder you blow the greater the cooling effect becomes. Many athletes are fully aware of this.

Can a fish really give an electric shock?

Yes. Electric fish – eels and ray fish for example – can generate voltages between the head and the tail of thousands of volts. They stun or kill their prey in this way. The current used may be as high as one ampere.

Biological chemical cells in the fish called electroplaques are activated by a nerve signal. The South American eel has one hundred and forty rows of electroplaques giving a total voltage of thousands of volts.

The fish does not kill itself with this voltage because of the way in which the rows are electrically connected. Being coupled in parallel ensures that a minimal current flows in the fish.

Is it possible for a person or an animal to make sounds that they cannot hear?

The simple answer is yes. A cricket for example, by rubbing its legs against its abdomen can produce sounds varying from seven thousand to one hundred thousand vibrations per second. It can however only hear sounds from one hundred to one thousand five hundred vibrations per second (or Hertz as they are called, after Heinrich Hertz). So it hears little of its own sound.

This is not a common feature. Most animals can hear much more of a frequency range than they produce. From a communications point of view it would be silly for them to make sounds they could not hear. It is also important for them, as protection, to pick up sounds they do not themselves make – particularly those produced by their predators.

Some examples are the following:

	Makes sounds of frequency (Hz)	Hears sounds of frequency (Hz)
Dog	452➔1,080 (vibrations/sec)	15➔50,000 much larger range
Cat	760➔1,520	60➔65,000 even wider range
Robin	2,000➔13,000	250➔21,000 narrower than dog or cat
Porpoise	7,000➔120,000 (like cricket)	150➔150,000 very wide range
Bat	similar to porpoise	
Man/woman	80➔1,200	16➔24,000 including those from a cricket that it cannot hear

For comparison, the top frequency of an average bass singer is 370 Hz; a tenor 440 Hz; a soprano 880 Hz; middle C 262 Hz.

When you skate, why do your skates glide along the ice rather than stick to the surface?

When the skates move they glide over a thin layer of water, the water being caused by the pressure exerted by the skater on the ice.

Pressure is force, in this case weight, applied to an area of surface. An average skater may weigh 800 Newtons if the mass is 80 kg or 176 lbs. This force is applied over the area of a skate, or perhaps a small area of a skate and the pressure is sufficient to turn ice into water at normal temperatures. It is interesting to note that for some time in the early 1940s women in stiletto heels were forbidden to board commercial aircraft, whereas massive African elephants were occasionally carried as air freight with no problem. The difference lay of course in the area of contact between traveller and aircraft. A 1,000 kg elephant has four feet making up approximately a square metre area in total. An 80 kg person might wear a stiletto heel of a few square millimetres dimension. As pressure is weight per unit area, the pressure exerted by a woman's heel is hundreds of times greater than that of an elephant. Similarly, in skating, only a small area of skate is in contact with ice so the pressure is great and the ice melts.

The pressure required to melt ice is highly temperature dependent.

On a cloudy day in late winter, towards the end of March, with apparently hardly any light filtering through the heavy sky, if an observer looks at a fresh snowbank she may be overcome by the brightness of the light that is reflected. How can such perfect whiteness be observed from a snowbank when it appears little light from the sun is reaching ground level and everything around is uniformly grey?

The most relevant fact in relation to this question is that new snow will perfectly reflect white light that is falling upon it, irrespective of the source. When you look at a cloud and you see it as dark grey, or even black, you see regions that are not transmitting direct sunlight. You are not aware of the light that is actually still being transmitted through the cloud and reaching ground level, and of course any light that reaches your snowbank is likely to be reflected almost perfectly by the crystals that are there. It should also be noted that the light coming through the cloud is diffuse and going in a whole variety of directions on transmission. The diffusely transmitted light from the whole cloud surface within the solid angle acceptable by the snowbank is however available to the snowbank and so one must not be fooled because a cloud looks dark or even when the whole sky looks grey, into thinking that little light is reaching the earth's surface. Indeed, the light is more than substantial under these circumstances, and although the listener's question gave the impression that the light reflected from the bank was intense, there was no suggestion that snow blindness would be involved, as might well be the case if one stared continuously at a snowbank when the full direct sunlight was falling upon it. Many skiers know that this can indeed be a problem and most people that engage in that sort of activity should wear some form of eye protection in the form of Polaroid sunglasses or filtering material. So, whereas the observation sounds dramatic and the components involved in it seem conflicting, there seems little doubt that what is observed is merely what a physicist would qualitatively expect.

The number of a certain house on Wellington Crescent is 1443. The number is displayed in large brass letters and it does not take a very observant person to see that whereas the two middle letters, that is to say the two 4s, are bright and shiny, the two numbers on the outside are tarnished, pitted, and rough despite the fact all four were originally erected at the same time. How can this be?

Perhaps rain falling on these letters from the lamp supports above is falling preferentially on the outer two letters rather than the centre ones and eventually, through acidic action, destroying those two. The main problem is that it has not rained in Winnipeg, in any significant way, for about five years and to depend on that explanation would seem to be flying in the face of reality. The second possibility is that the prevailing wind from west to east is blowing dirt from the prairie or from the yard across the surface of the four numbers and preferentially cleaning, as in sandblasting, some numbers more than others. Had that, of course, been the case, the number 1 would have been most shiny and the others less so, as one moved to the right and from 1 through the two 4s to the 3. This is not in accordance with the facts of the question.

The answer is that the two 4s are from a good batch of numbers in which the alloy was of quality; the 1 and the 3 however are not really of brass composition at all. They are a mixture of copper and zinc that is not within the proportions of alpha brass or perhaps even beta brass, for that matter. Up to thirty-six per cent in proportion of zinc is found in a quality brass. For proportions larger than that and up to forty-five per cent, you will have a brass that is less ductile but stronger. But beyond this, and also for a much lower proportion of zinc, you will not be dealing with brass but with a mixture of copper and zinc in unusual proportions. Once you have separate pockets or small cells of copper and zinc randomly displaced throughout the body of the material, the possibility of chemical action becomes a real one. You can have little galvanic cells dotted throughout the surface which, if joined by some piece of debris or conducting material in the form of dust or something else, may actually, through chemical action, generate a cell that will, through electrical activity, generate holes or pitting in the material. The surface of the numbers 1 and 3 is certainly consistent with such a happening and the overall roughness of the surface which was originally smooth may be attributable to electro-chemical action.

Why is it difficult to make snowballs at very cold temperatures?

Ice cubes in a bowl, like snow crystals in the environment, can bind together through the thin layer of water that the pressure of one on the other creates at below-freezing temperatures. When water molecules are between two icy surfaces more of them leave the water and become ice, than leave ice and become water. In this way bonds form between snow crystals or ice cubes which bind them together. This is how snowballs are made. However in very low humidity and when water has been frozen out, e.g. at -20°C, it becomes impossible to make snowballs. Snow becomes a fine talcum-like powder.

I observed last week that lamps in a car park had bright plumes of light visible above and below them. It was -28°C at the time and the additional light was clearly seen. What causes this?

In still cold weather, the ice crystals which form are often flat and six-sided – like a pencil with which a student writes. As the ice crystals fall, they may project one or more of the six horizontal surfaces to the eye. An observer, looking at the light from a lamp in a car park, may see either direct light or light reflected from observable faces of the pencil-like crystals in the air. Light is often reflected from crystals high above or sometimes below the lamp. The result is a plume through the lamp, and has been reproduced by computer from an analysis of this kind.

Why do Inuit people, in very cold weather, turn their fur coats inside out?

Somewhat like polystyrene foam, fur always contains pockets of air in it. If the fur is on the outside of the coat, conductivity heats the trapped air; it then expands and is lost only to be replaced by cold outer air. When the fur is inside the insulating leather of the coat, body heat is trapped and retained, so remember that in January!

Part Two
Theoretical

Assume the earth to be a perfect sphere and that it is possible to tie a string around it so tightly that it is impossible even to push a sheet of paper between the string and the earth. You now cut the string and insert an additional twenty metres of string into the complete circle that is girding the earth. In the new configuration, will it be possible to (a) slide a piece of paper between the string and the earth, (b) crawl below it, or (c) walk underneath the string without touching it? It is not necessary to know the circumference of the earth in order to solve this problem.

The solution to this problem lies in the nature of pi (π). This symbol has got an unusual significance, being defined as the ratio of the circumference to the diameter of a circle. The quantity has been measured with increasing accuracy over many centuries since the Greeks first interested themselves in the relationship. The value of π is 3.14159265, etc. continuing, as far as we know, forever and being designated as an irrational constant. It is currently possible to measure π to several millions of decimal places and research on the number still continues. Pi, therefore, is slightly greater than 3.14 and much less than 3.15 and this fact is all that one needs to know in order to solve the problem.

When we add twenty metres to the circumference of the given circle it follows from the nature of π that we thereby increase the diameter of the string surrounding the earth, in proportion to the increase in the length of the circumference. So, if the circumference is increased by twenty metres, the diameter of the string circle will be increased by twenty metres divided by π. If we now take the larger number, 3.15, for pi and divide it into twenty, we find that the diameter of the string must have been increased by about 6.35 metres. Remembering that the string is equally placed around the earth in the nature of a halo, this means that where our observer stands the string will now be just approximately 3.2 (half of 6.35) metres above the earth and be similarly placed at all points around the circle. At a height of approximately 3.2 metres, it should be possible for even the tallest basketball player in the world to walk effortlessly underneath the string in its new and higher position.

It seems counterintuitive that such a small increase in length of string can have such a profound effect as the circumference of the earth is around forty thousand kilometres.

What is exponential growth and how does it relate to the world economy?

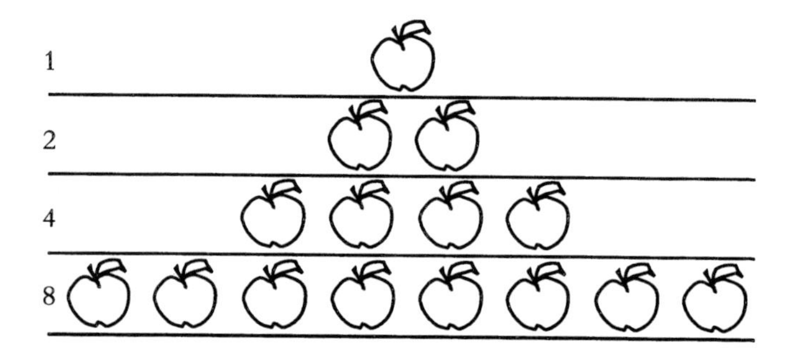

Exponential growth is the regular increase in a product or quantity that is characterised by a doubling time. For example, suppose I eat one apple on Monday, two on Tuesday, four on Wednesday and eight on Thursday – my appetite for apples is showing exponential growth. But this growth can not be sustained. By Friday, I would be eating sixteen apples and by Saturday, thirty-two. Normal bodily functions determine that such growth must level off or decline in the not too distant future – certainly, it could not continue forever. Also, when we talk about exponential economic growth, again we think of a doubling of specific quantities – if not in days, but certainly in years or decades. Population and energy consumption are two such examples.

First consider population growth. Worldwide, population still seems to double regularly in an exponential fashion – but for some individual developed countries, growth already seems to be at a standstill.

Secondly, there are developing countries whose economic growth – in terms of its energy consumption – seem to be increasing exponentially. This occurs as poorer countries strive to become industrialised and raise their standard of living to that of their developed neighbours.

Finally, there are countries like Britain, Germany and the United States which have levelled off to the point where their energy consumption per head is now constant year by year. In Britain, there has been no increase in energy consumption per person per year in the past one hundred years.

Some researchers have suggested that internationally, further economic growth should be halted – that this planet merely make the best of what it already has. That would have disastrous political implications for the earth as a whole. There would be no way in which the aspirations of the developing world could be realised. Nor could standards of living be raised to those of already developed prosperous partners. It seems much more appropriate for the developed world to encourage countries such as India and Ethiopia to adopt the best method to accomplish economic growth and financial independence. Environmentally acceptable and essentially pollution-free energy sources are available to do the job. After all, burning sticks never was nor can ever be, pollution free on the large scale. Temporary exponential growth in the production of energy can rapidly improve the standard of living of developing countries. If history tells us anything, these countries can reach the plateau attained by Britain a century ago. In the process, they can increase their wealth, solve the difficult problem of overpopulation and arrive at the desired results of constant energy consumption and a stable population.

How is it possible for a living organism to be bigger than a beetle? Why does it not collapse under its own weight?

The problem with a large object is that the surface area grows more slowly in relation to its volume in overall growth. Take for example a spherical object of given radius. If that radius is increased by a factor of ten, as in blowing up a balloon for example, the surface area increases by a factor of one hundred, and the volume by one thousand. If the object is not a balloon but a solid object like a panda bear, then an increase of one hundred in surface area with growth will not only involve a one thousand times increase in its volume, but a similar increase in its weight, making survival much more difficult for a variety of reasons. The first is that most functions of the animal are served by its surface. A worm for example is almost all surface! Digested food reaches the body and oxygen is absorbed through this interface. Secondly, with increasing weight, the strength of bone to support the animal is greater and depends on bone cross-section. For a small animal like a beetle this is unimportant, but because a cross-section increases as the square of length, and a volume as the cube, bone size becomes vital to the support of larger animals. So, how do large animals deal with the problem of inadequate surface area? They do as humans do; they develop internal organs and thereby generate internal surfaces such as the lung (a bag of surface for gaseous exchange), a circulatory system (which moves material to parts of the body not reachable by diffusion), and villi (hairs which increase the surface area of the small intestine for absorption of food).

Small animals do not need these sophistications. They basically breathe through hollows in their outer surface. For large animals or humans, however, such hollows would have to be so deep and numerous that the organism would not only fall apart, but have no room for internal organs. Being bigger than a beetle clearly complicates life immensely.

Why do grass and low-lying plant life get so wet at night even in spring and early summer?

Dew is formed on surfaces which are at a temperature below 'dew point', the point at which dew forms. Dew consists of fine droplets of water that form on solid surfaces because the surface (e.g. grass or leaves of low-lying plants) is at a temperature higher than freezing point and lower than the dew point of surface air. At such a temperature water vapour condenses out on the cold grass or leaves, because the air is more than saturated with water vapour.

Of course, heat from the subsoil is conducted to the base of the grass or plant stem. However the plant stem is a good heat insulator and inhibits flow of heat from below. Also air trapped between leaves and grasses performs the same function.

Nonetheless the upper grass blades become cold enough to fall below the dew point of surrounding air because plants transpire water vapour and cool. They also directly radiate heat into the sky. Under these conditions grass temperature can easily fall below 'dew point'.

What is the greenhouse effect and does it relate to global warming?

Insofar as the so-called 'greenhouse effect' relates to the absorption of heat (infra-red radiation) by gases, then global warming should be a perfect example. What is not clear is what this has to do with 'greenhouses'.

Greenhouses are basically structures which trap air. Solar radiation is absorbed by soil which then heats air in contact with it. If it were otherwise air heated by direct solar radiation would be warmer than lower air – which is not the case. Only carbon dioxide, ozone, methane and water vapour absorb infra-red radiation. *Oxygen and nitrogen do not.* Incidentally, carbon dioxide is present at only three hundred and forty parts per million in air. Water vapour is the principal greenhouse gas.

So in reality greenhouses are shelters from the wind. They suppress heat transfer by convection to the outside. In January 1961, the minimum temperature inside a small greenhouse was recorded as 1°–2° Celsius colder than outside. The result was published in the *Journal of Applied Metrology*. The most important features of a greenhouse are thick glass and a sound structure for trapping air.

Global warming requires more than a greenhouse analogy for its explanation.

Why is a thread of glass so strong?

The fact that a fibre of glass less than a thousandth of an inch thick is stronger than steel is not because it is thin. It is because it is smooth. The reason why most materials are weak or brittle is because of cracks. Bricks, iron and graphite can all break because of voids or cracks in the material. Some cracks may only be of atomic dimensions. Any material can be kept strong if it is free from cracks.

How much energy per day does the body really need and what is it used for?

The power requirements of the average human body in repose are equivalent to that of a 100 watt light bulb or 100 joules of energy per second. In order to function properly, that amount of energy must be supplied to the body on a regular basis through consumption of nutritious food. Today, almost every country in the world except for the United States, Burma and North Borneo, have adopted the International Metric System, of which the joule is the unit of energy. However, in earlier times the unit of heat energy was the 'calorie', equivalent to 4.2 joules of mechanical energy. As this unit was so small, nutritionists invented a 'great' calorie equal to 1,000 little calories, or 1 kilocalorie. A power of 100 watts or its equivalent, 100 joules of energy per second, is equivalent, in one day, to 2,000 calories, the number nutritionists most often quote as our basic energy requirement. The body is therefore equivalent to a 100 watt light bulb except for the fact that nutritional energy rather than electrical energy is involved. Energy is what makes it possible to do work. Food in the body is turned into chemical energy and allows us to operate our bodily systems and incidentally to maintain our temperature at 37°C.

Energy of one kind is regularly turned into another. When we 'run' or 'pump iron' muscles get tired. That is because we synthesise lactic acid in the muscles and bodily energy is in a sense wasted doing that. If we are very active however and do not supply enough food for the body's biochemistry, we will feel tired because the energy input is just not there.

I make two drinks in identical glasses – one is rye and soda water, the other rum and Diet Coke. The volume of each is the same as the other. Each glass has three ice cubes from the same tray added. The ice in the rye and soda water glass however melts twenty minutes prior to that in the rum and Diet Coke. Why?

The answer has little to do with the different sweetness (sugar content) of the two drinks.

Sugar, of course, can cool a drink as it dissolves, as is the case with coffee, but the effect is small. The sugar level should not give a noticeable effect. Salt, of course, in significant quantities can lower melting points – in the case of water, ice forms at 22°C lower temperature when a salt and water mixture is involved. Salt however is not mentioned as an ingredient of either soda water or Coke and this should therefore be a negligible effect.

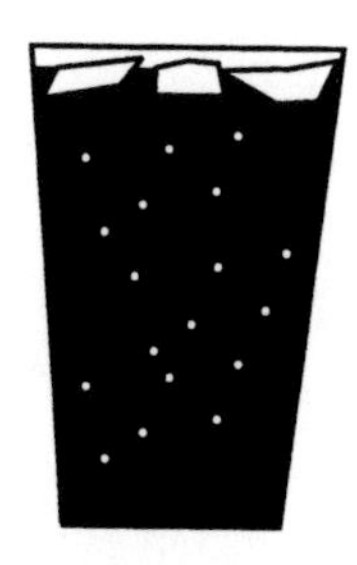

A possible explanation is that carbon dioxide gas which gives soft drinks their fizz is more soluble in Coke than in water. Under pressure, soda water contains a significant amount of the gas but carbon dioxide is not readily soluble in water. Coke seems different. It has a higher solubility for the gas and does not go 'flat' in air so readily. In addition, it is my belief that the viscosity of the cola drink is large enough to inhibit the flow of bubbles and heat to the surface. Once a soda water bottle is opened, gas bubbles rush to the surface and into the air once the pressure is released. The gas bubbles may carry heat away from the body of the liquid very efficiently (hot gas rises) in the case of soda water – less so with Coke. Now, as the surface of liquid in the rye and water glass is almost covered with ice, the heat from the bubbles will be absorbed by the ice eventually melting it more efficiently than if only the external surroundings were providing heat. The presence of alcohol in these drinks does not seem to be a relevant factor. Indeed rum and Coke should behave exactly as rye and Coke; rye and soda as rum and soda. Someone with a non-discerning palate should test that. Perhaps Coke and ice, and soda water and ice should also be compared, if this hypothesis is correct. I leave it to the reader (drinker) to investigate the possibility of difference there.

Is there really wine that talks, as a scientist recently suggested on radio?

Not literally, but it tells a tale to those skilled or wise enough to listen. Scientists have discovered that it is in fact the water rather than the wine that gives the show away. Water can be used to identify the precise locale from which the grapes originate.

It transpires that the ratio of two types of oxygen present in the water in grapes varies markedly with the region from which the grapes are picked. Ordinary water is H_2O. Two atoms H to one of O as every school kid knows. But oxygen atoms come in three kinds ^{16}O, ^{17}O, ^{18}O – all stable. Ordinary oxygen in air is 99.75% ^{16}O, .205% ^{18}O, .039% ^{17}O. The ratio for ^{18}O to ^{16}O varies over the earth's surface and so H_2O may have more or less of ^{18}O while being chemically the same.

In general, the ^{18}O to ^{16}O ratio in rain/precipitation from different areas shows a graded decrease as we go from the equator to the poles. Indeed, the distance from the sea, and the height above sea level affect the ratio significantly. This ratio in Germany decreases from the north coast to the southern mountains.

The River Rhine is supplied by precipitation from southern Germany, so water near the banks of the Rhine shows lower ratios than that near the bank of the Moselle which represents precipitation in the highlands of middle Europe. In the north, the influence of evaporated sea water also affects the ratio.

In general the temperature at which chemical processes take place is important. The $^{18}O/^{16}O$ ratio in leaf water is temperature dependent – so therefore the distance from the equator is important to this ratio, which is determined by the whole climate.

Basically then, wine comes from grapes. Grapes have a large volume to surface ratio – loss of water is therefore a slow process. Grapes store an oxygen ratio higher than local rain or ground water. In evaporation, $H_2{}^{16}O$ which is lighter emerges first and $H_2{}^{18}O$ (being heavier) afterwards. This enhances the $^{18}O/^{16}O$ ratio. When wine is made, the history of the water in the grape is therefore preserved.

A compass points north in the northern hemisphere and points towards a magnetic pole. What does it do in the southern hemisphere? Is there a magnetic pole there, and does it point towards it? What happens at the equator?

A compass consists of a permanent magnet set so that it can point in a particular direction. Magnetic north as it is called is the direction in which the North Pole of a compass magnet is set to point.

It was the physicist William Gilbert (1544–1603) who first realised that the earth behaved as a two-pole magnet. It was also confirmed from studies of permanent magnets that unlike poles attract each other, and like ones repel. Therefore the 'north' pole of the compass, in seeking north will be attracted by the 'south' magnetic pole of the earth. The magnetic pole situated in the region of magnetic north is therefore the 'south' pole of the earth as a magnet. The magnetic poles of the earth however do not correspond exactly with the geographic poles. Indeed magnetic north to which the compass points is approximately 11.3° in angle from geographic north, and barely within Canada.

Currently the magnetic north is seemingly on the move, and predictions are that within a century it will have left Canada on its way toward Greenland. The answer to the other parts of the question follow from the general remarks about the earth as a magnet. A compass needle whether in the northern or southern hemisphere, or at the equator for that matter will always point towards magnetic north, irrespective of where it is locally positioned.

Why do the earliest sunrise and the latest sunset not occur on the same day?

As measured by clock time, the earliest sunrise and latest sunset of the year do not occur at the summer solstice nor does the latest sunrise and earliest sunset occur on 21st December at the winter solstice.

Certainly in the northern hemisphere the shortest day is 21st December and the longest 21st June. The earth's equatorial plane is tilted twenty-three degrees twenty-seven minutes to the plane of its orbit round the sun, and for a few days around these times of year the maximum and minimum altitudes of the sun at noon vary little. Indeed the sun seems to stand still, hence the name solstice. In Winnipeg, because of its latitude of ~49° north, the high and low angles of the sun are ~73½° and ~26½°. As 21st December is the shortest day we might expect the latest sunrise and earliest sunset to occur on that day, but in fact the sun continues to rise later each day and on 24th December, Christmas Eve, it rises two minutes later than on 21st. Also the earliest sunset is several days before 21st. The opposite situation occurs in summer.

The absence of symmetry here has worried many scientists. But astrophysicists tell us it is because of the 'equation of time'.

The earth travels round the sun in an elliptical orbit – not quite a circle – and speeds up as it gets nearer the sun, slowing down when further away. Its greatest speed is in early January and least in early July. As a result, as observed from earth, the sun moves at different speeds eastward across the sky relative to fixed stars. This is different to the east-west daily apparent solar movement. So there is a difference between sun and clock time.

On the equinoxes – equal periods of day and night, 21st March and 23rd September – the noon height of the sun changes rapidly from day to day. So its progress eastwards is less than in December and June. The change in speed of the earth in its orbit means that it is impossible to replicate its behaviour by mechanical means – e.g. a clock. So in practice the real sun is replaced by a pseudo-mechanical sun which moves at a regular and constant rate and keeps what is known as Greenwich Mean (solar) Time. Sometimes clocks are ahead of solar time (equinoxes) and sometimes behind.

A correctly mounted sundial will not agree with a clock as it measures the position of the real sun. Sometimes it will be ahead, sometimes behind. The difference between clock and sundial time has the grand name of 'the equation of time', and this difference gives rise to the problem identified in the question.

Is it true that Antarctica contains ten times as much ice as the Arctic? If so, why?

Antarctica is in fact the world's fifth largest continent. It is surrounded by a stormy icy sea – the Antarctic Ocean. In size it is 13,209,000 sq. km covering one-tenth of the earth's surface. The average height is 1,830 m, higher than any other continent. The ice and snow has an average thickness of 2,440 m. Average temperatures are much colder than the Arctic – making it the coldest region in the world.

The Arctic, on the other hand, is an ice covered ocean between North America and Eurasia. The temperature of the ocean water is around -1.7°C – the freezing point of partly salted water (at 32 ppt*). Antarctica has eight times the ice cover of the Arctic Ocean.

The reason is this: the heat capacity of water is great. It is hard to heat and slow to cool. Land is a poor heat conserver and radiates heat away rapidly, allowing ice to form readily. Arctic ice is on top of an ocean. Here the water takes a long time to heat but once it does it holds on to its heat well. The Arctic 'stores' summer heat and lives off its savings in the winter.

What time is it at the North Pole?

The simple answer is: what time would you like it to be?

At the North Pole we have what the mathematicians call a singularity. Precisely at the pole there is the point where all the appropriate meridians (lines [circles] through both poles) intersect and all time zones come together. Exactly at the pole it is a matter of mutual agreement as to what time it is.

The idea of time zones is due to Sir Sanford Fleming – surveyor with the Canadian Pacific Railway. He advocated a

*Parts per thousand.

world system of keeping time. The railway had made the old system obsolete where each area set its clocks locally according to astronomical conditions. In 1884 the system of international standard time was adopted.

Time zones are measured from a meridian passing through Greenwich (London) defined as 0° longitude. From Greenwich every 15° of angle is a basic time zone of constant time, but each successive zone is one hour different in time. Going from west to east, hours are added until at the International Date Line a day is lost.

Specifically, time is the distance between events, and can be measured anywhere. At the poles all meridians meet. If you are positioned at a pole you can walk in a circle and momentarily have a foot in any time zone. Whether you start at 0° (Greenwich) or 180° (Date line) is a matter of choice. The important thing is that every twenty-four hours the sun (if visible) will be in the same position as when you first observed it.

Indeed if you put a flat sundial at the North Pole, the shadow will move 15° every hour until full circle is reached – so twenty-four hours is always twenty-four hours. Whether you call your starting point midnight or noon is incidental.

Polar bears probably don't care!

Part Three

Home and Kitchen

When a thick soup or broth is stirred vigorously, the rotation of the surface occasionally dies out quickly to be replaced for a few seconds by rotation of the soup in the opposite direction. Can you explain what is happening here?

This phenomenon is an interesting example of the elastic recovery of a visco-elastic fluid. Let me explain these terms. In an elastic solid, such as a spring, mechanical energy given to it is returned rather than lost. Treacle, on the other hand, is a viscous fluid; it tends to resist flow. Some liquids actually combine both properties, elasticity and viscosity. Certain jellies, soups and doughs are examples.

The swirling soup will slow down because of friction with the sides and bottom of the saucepan and between the layers of soup in the pan. The rotating soup on the surface is the last to cease motion but is then pulled back by an elastic force between it and the rest of the soup in the pot. This is very like a spring being compressed and then released. You can also see this phenomenon in other situations. If you start to pour out a visco-elastic liquid, and stop suddenly, the liquid then breaks and the stream behind the break rebounds toward its original container. Interesting stuff.

In soap powder and detergent advertisements we hear that clothes can be much 'whiter than white'. We even hear talk of a 'blue whitener'! Can something be 'whiter than white', and what can 'blue' have to do with it?

'White light' in the physical sense is light containing all colours of light in equal intensities. In practice the term is also used to describe a wide range of distributions of colours with different intensities. Also, a white surface is one which reflects all colours equally well. In practice no surface is perfectly white.

When normal sunlight falls on an object the intensities of colours at the ends of the visible solar spectrum are usually much less than those in the centre, yellow in particular. However, addition of a fluorescent powder to a supposedly white material can instantaneously transform invisible ultraviolet rays from the sun into visible violet or blue, thus adding both to the total illumination of a shirt or blouse, and to the intensity of blue light in the reflected spectrum.

Invisible ultraviolet light is more energetic than visible blue light. Therefore, when falling on a fluorescent surface it loses some small amount of energy and reappears as a less energetic but visible (usually blue) light. The fluorescent additive is a 'blue whitener', and the shirt or blouse being cleaned is now 'whiter than white' or 'bluer than white' for the reasons given.

How can we tell whether an egg is boiled or not, without breaking the shell or using any additional equipment?

The fact is that a boiled egg spins differently to a raw one. Take your egg, place it on a flat surface and twirl it! A cooked egg will revolve much faster and longer than a raw one – indeed it is difficult to make the raw egg turn whereas the boiled one, if flicked hard enough, may even stand up on its narrow end.

The explanation is that a hard-boiled egg is a rigid body and turns as a whole. The raw egg has liquid contents which retard motion by the force of inertia and act as a brake on the rotating object.

The oil in a thin frying pan will heat faster than in a heavier iron pan, but the heavier pan will do a better cooking job. Why?

The heat capacity of the thin pan is much less than the heavier one. However, heat is transferred from the stove to the food through the pan in each case. Once the food is placed in the frying pan with low heat capacity it removes heat from the pan and momentarily lowers its temperature and that of the oil. As a result the heat does not sear the food, which instead absorbs more oil. Moisture is released from the food which being in contact with the heat source then sticks to the pan. The heavier pan has a large heat capacity and barely changes temperature on the deposit of food, which is then heated uniformly and cooks well.

Why do leaves in a cup of tea collect in the centre after stirring?

When tea is rotating round the centre, the acceleration for such circular motion comes from the pressure difference between fast-moving tea near the wall and tea near the centre. The velocity at the wall is highest. However it is a secondary effect that deposits the leaves in the centre, due to friction with the bottom of the cup. The bottom layer of water in the teacup is slowed by rubbing against the porcelain or glass bottom. The speed of rotation of the outer liquid is less than at the surface. Therefore a pressure difference exists between top and bottom layers as well as between centre and rim. As a result, a sample of tea originally at the outer part of the top surface not only circulates around the centre but descends along the wall to the bottom because of the pressure difference between outer top and outer bottom. To replace fluid and leaves flowing to the bottom there is a matching flow of liquid up the central axis. So while tea is circling it is also flowing from outer top to outer bottom to centre bottom to centre top. Tea leaves on the bottom are captured by this secondary current and brought to the surface.

Can plasticisers used in plastic cling wraps move into the food they contain during microwave cooking?

Yes they can. Plasticisers are organic liquids with high boiling point and low volatility that are mixed with polymers to make them into very flexible film. As food and a covering film are heated, plasticiser can migrate into the food. Some organic compounds are toxic and carcinogenic, so eliminating this possibility is important. Even defrosting a chop in film wrap is a potentially unwise act. It is also unnecessary. It is also my understanding that no government agency in the US or Canada tests 'microwave-safe' products to prove that plasticisers do not enter the food. In particular, do not use margarine or yoghurt containers to heat food in a microwave oven. These have not been tested for chemical migration and are susceptible to chemical breakdown if heated. Use glass or heat resistant ceramic. Always transfer food from the container it comes in to appropriate cookware before cooking in any oven, conventional or microwave. Not all plastics or plastic films contain plasticisers, but most do. It always pays to be careful.

Why does frost not form on chicken placed in a freezer in a brown paper bag?

Frost is frozen water vapour or moisture. Vapour from a chilled product will usually be removed by sublimation or evaporation in a process related to freeze-drying. When a plastic or other freezer wrap is used, moisture is held inside the package and its presence reduces or inhibits freezer-burn, a feature of over-dried frozen food. On the other hand, paper is a finely woven fabric in texture; it is porous and therefore gas or vapour will pass through it. As a result, no moisture is left behind to freeze on the product.

Moisture normally recondenses from vapour on the coldest surface in the freezer, usually near the cooling coils. In modern refrigerators a heat pump is added to inhibit the recondensation, in which case defrosting of the freezer is never needed.

Why is the whipping of egg whites in a copper bowl always successful whereas the beating of egg whites in a glass bowl is usually a complete failure?

All whipped egg whites consist of a film of bubbles whose walls contain a film of protein long-chain molecules which are an important part of the human diet. Laboratory experiments have shown that whipping egg whites in a glass bowl results in a cooking film that is so flimsy that the bubbles are likely to burst immediately. The surface tension, in fact, is too low. With the copper bowl, however, minute trace amounts of metal are absorbed as a result of the whipping action. The copper atoms from the bowl form a chemical compound with protein albumin in the egg white. The result is a stable surface of bubbles which not only looks great but is a triumph for the cook. It is therefore the addition of copper in such minute quantities that it in no way becomes toxic even to egg white aficionados that makes all the difference between a successful whipping and an unsuccessful one. Without the metal additive, the whipping process always works imperfectly.

A well-known female broadcaster wants to know the reason why scum forms frequently on tea.

Scum is any layer of impure matter on the surface of a liquid. It does not form on tea infused in pure distilled water, but generally it is present to some degree otherwise. Researchers at Imperial College London have shown the phenomenon to be due to the hardness of the water used. Analysis of the scum shows it to be largely calcium carbonate formed by a chemical reaction between water and tea. The water must however be 'hard'.

Fresh water contains much less salt than sea water. However it may contain dissolved material that makes it unsuitable for many uses. Water containing ions of calcium, magnesium and iron is called hard water. Hard water makes washing with soap difficult because these ions form an unpleasant scum with soap. Salts of calcium tend to come out of solution and form scale on kettles, pipes, etc. Water softeners are used to get rid of calcium carbonate and magnesium hydroxide. They simply consist of lime and soda ash as an additive.

In order to solve the problem the addition of an acid will reduce the amount of scum; whereas the addition of an alkali will make it worse. Citric acid in the form of slices of lemon will often remove the problem completely. The calcium carbonate is broken down and the scum is lost. On the other hand, the addition of milk makes it worse as it is an alkaline product. Now, if you don't want to add lemon to your tea, do not leave the teabag too long in the cup, as the scum increases with time. Also if tea is kept warm once made, scum increases with time, so drink it quickly! Or keep your eyes closed!

When I put a spoonful of instant coffee grains in water just below boiling point, the water seems to start boiling. Why would this be?

Certain solid substances attract gas molecules to their surface – this is called adsorption. Activated charcoal for example is a substance which removes large amounts of poisonous gases or impurities from air, and is used in gas masks, cigarette filters and aquarium cleaners. Coffee grains are similar and adsorb gases on the surface. On heating much of the gas is removed and bubbles of gas are seen. The coffee is not really boiling – just bubbling.

In each case gas molecules are attracted by those already on the surface. The larger the surface area, the larger the amount of gas adsorbed. Activated charcoal is carbon full of holes, with many kilometres of surface in the holes in the material. It therefore attracts gas molecules and holds on to them. Coffee grains also are porous and have a large surface area. However, adsorption decreases with higher temperature, so when coffee grains are placed in hot water they release much of the adsorbed gas as bubbles and simulate boiling. In the case of coffee much of the released gas is air which leaves the liquid. In most cases, the coffee grounds have adsorbed gas from their earlier production environment which finally ends up in your coffee.

Why does food cooked in a microwave cool faster than that in a conventional oven?

Microwave ovens cook by absorbing microwaves (radio waves of approximately 12 cm in length). The amount of absorption of energy with depth depends on the frequency of the microwaves. Most microwave ovens have a frequency of 2,450 MHz and invert water molecules 2,450 million times a second. At this frequency the first 2 cms of meat absorb most of the energy. If the piece of meat is small, heat from all directions will flow from the surface to the centre and the centre may be hotter than the outside. However for a large piece of meat or fowl, the outside is likely to be much warmer than the inside and will cool as heat is conducted to the centre after removal from the oven. Also if the frequency is higher than that suggested, the depth of penetration may be less.

A conventional oven, on the other hand, usually supplies heat evenly leaving the contents of the cooked product uniformly at the temperature of the oven. So, in a microwave oven, irrespective of surface temperature, the centre of the product may rapidly remove heat from the surface and spread it through the meat as it sits. The conventional oven-cooked product is usually uniformly at the same temperature so heat does not flow from one part to another.

Why should you never cook salty food in a microwave oven?

Reports in *The Lancet* and elsewhere have indicated that bacteria in food, particularly salmonella and listeria are not being destroyed by cooking or reheating in a microwave oven. When it is recalled that microwave ovens are marketed as 'heating from the inside out' a recent paper in *Nature* puts in perspective the fact that no scientific data supporting this fact were ever made available. Indeed it transpires that temperatures in food may not be sufficient in many cases to kill the bacteria present. The presence of salt however plays a major role in this result.

Researchers in Leeds University recently took ten different pre-cooked TV dinners and added salmonella and listeria bacteria to them. They used ten different makes of Government-approved microwave ovens and heated the food for the times specified on the packets. They found in all cases that approximately half of all bacteria survived – a disturbing if not alarming fact.

Strangely enough, previous trials successfully completed by Government scientists were all performed on 'unsalted mashed potato' whereas TV dinners and most foods normally eaten contain salt in varying quantities. The Leeds scientists thought salt might be the key. They took samples of mashed potato with different amounts of salt added, either common salt, monosodium glutamate, ammonium chloride or potassium chloride. They measured 'core' temperatures in the samples for different concentrations, either zero, less or greater that those found in preheated foods. The temperature change after one minute heating varied from 30°C for unsalted mashed potato to 3°C for the sample with most salt. The surface in all cases was however hot.

It seems that instead of working on the water molecules (polar) in food and heating the middle of each sample through friction (from the inside out), when salts are present they form positive and negative ions which flow as currents (electrical) on the surface of the food, taking energy from the microwaves that should be used in heating. The more ions, the less heat to the core so that only the surface boils. Therefore, in cooking food in a microwave oven, cook 'fresh' food without salt or preservatives, rather than pre-cooked foods with salts already added. Manufacturers of pre-cooked foods often use preservatives, and these hold the seeds of potential disaster.

My garden tomatoes will not ripen and instead stay green. If I take them off the plant and put them in a brown bag or drawer they ripen eventually. Why is this?

This is an interdisciplinary problem. Growing of tomatoes requires adequate climatic conditions – warmth and a suitable length of summer for example. In many countries tomatoes are only grown in greenhouses. This summer in Canada the fruit has refused to ripen outdoors so other steps must be taken.

The solution is that putting the plants indoors causes chlorophyll degradation and the loss of green colour. A ripening hormone – ethylene – present in most fruits then reddens the tomato – just as ethylene gas is used to make green oranges, orange, artificially.

Indeed if a ripe banana is placed in with unripe tomatoes it emits so much of the ethylene gas naturally that ripening occurs more quickly.

I wish to report the apparent explosion of an empty orange juice glass on my kitchen counter at breakfast time. How could this have occurred? It seems that nothing struck it nor was it on a vibrating surface. It broke into a myriad of small pieces. What can be the explanation?

Glasses do not often explode or implode. However a second questioner recently experienced a similar phenomenon. His 'cornflakes' dish, sitting alone on a shelf, suddenly burst apart showering the area with small pieces of glass. So, clearly what the questioner describes is not a totally unusual event.

Excessive stress in the material must be responsible in each case. When a glass is made and the material cools and hardens, stresses are often set up in the material. Recent popular designs in cheap glass – often eight-sided or angular designs – seem to have stresses contained inside them which even a ray of sunlight may cause to erupt into action. Everyone knows that automobile windows can shatter into small fragments on damage. These are pre-stressed for safety reasons. However the result is the same.

Stresses occur in materials when glass that is formed into unusual shapes sets by cooling. If cuts and angles occur in the glass and it is naturally brittle or fragile, rupture of the glass may suddenly occur.

Something as simple as a few degrees' difference in temperature from top to bottom or inside to outside may do it. A hot dish placed on a cool surface may lose heat rapidly from the point of contact and cause thermal shock elsewhere. A little sunlight or vibration from a passing vehicle may do it.

In order to tell if a glass is likely to explode take two Polaroid sunglasses, place them one behind the other, and rotate one with respect to the other so that the view through them is dark. If you now put a glass or a windshield between the two a pattern of light may appear, indicating stress forces in the material. A study of cheap octagonal glasses might indicate several to be under stress.

The problem usually arises only in the case of mass-produced cheaper and fragile glasses. Lead crystal glasses for example seem to be of such quality as to limit such explosions. The cuts in the glass are obvious and deep but all stresses seem adequately contained.

Occasionally, during the night, I get up and go down to the kitchen, open the fridge, and take out a canister of yoghurt. Sometimes I notice that if it is a new carton and a spoonful of the substance is removed cleanly from the centre of the carton, the hollow left behind in the substance will have filled up with liquid by the time I return in the morning for some additional sustenance. Why has this happened and where has the liquid come from?

It is clear that thixotropy of some kind is involved. Technically, thixotropy is a structural property that causes certain gels to liquefy and then to solidify again. The breakdown to the liquid state is a function of time and shear rate. In general, solidity returns over time in the absence of shear. Once a shear force is applied to a substance that is only marginally solid, in this case by a spoon, the yoghurt is enabled to flow like a liquid while the shear force is being applied. This is known from the behaviour of such materials as margarine, which is solid and only flows when it is actually being spread; tomato ketchup, which has some of this property; and many ceiling paints which consist of a gel which flows only when the paint is being applied. For those not too familiar with the idea of a shear force, a shear force is applied by hedging shears when a garden hedge is trimmed. In respect of the actual question, the angle of attack of the spoon generated a shear force followed by liquid flow. Once the hollow was filled with liquid and the shear force removed, the liquid solidified again.

Why do my eyeglasses get coated with oil on the inside when cooking french fries on the stove?

It is an interesting fact that whenever oil from a pan is turned into aerosol particles which then fly up under thermal currents past the face of the cook, there is little time for minute particles to attach themselves to the glass of the spectacles. On the other hand, if the spectacles are fairly close to the face, particles of oil may occasionally get trapped behind the eyeglasses and find themselves in semi-static or stagnant air, that is between the face and the glass itself. In this case, such small particles of oil may actually move around in random fashion and tend to coalesce into larger droplets, most of which will tend to migrate toward the coldest surface and condense further there. The inside of the spectacles would seem to be the most likely point to collect aggregates of oil droplets and, if this is indeed the explanation, then we do understand the phenomenon which is described by the questioner.

After preparing my hot white sauce in the usual way tonight, I poured it over a hot dessert allowing it to cool briefly. On inserting a spoon, the sauce then apparently exploded, covering ceiling and wall with sauce fragments. How could this have happened?

The answer is probably this: cornstarch, the essential ingredient of white sauce, is basically a visco-elastic material. In water solution it can be an elastic solid when mixed in a one-to-one ratio. The solution will withstand hammer blows from above while flowing thickly in the horizontal plane. Indeed, pouring a one-to-one mixture of cornstarch and water down the sink can result in perfect blocking if any further evaporation of the mixture occurs. The relevance of this fact to white sauce is simply that at a ratio of one part cornstarch to six of water, a palatable sauce is generated. At lower ratio something between a delicious liquid and an elastic solid can be achieved. If a sauce rich in cornstarch is prepared and cooled the elastic solid may compress air below it and on puncturing produce a significant explosion once the pressure is released. This mechanism after all is similar to the generation of sea spray when the crest of a wave falls into a preceding trough. The moral to this story seems to be – go easy on the cornstarch and all will be well.

I make muffins for my family in large batches. Occasionally I have mixture left over and fill only half of the bowls on the baking tray. When this occurs I fill the otherwise empty places with ordinary water for balancing purposes. After cooking at 240° for the appropriate time I find the muffins perfectly cooked, but little, if any, of the water in the other bowls has evaporated. How can that be?

The answer lies in the large heat capacity of water. The specific heat, and latent heat of the liquid are both high. As a result the heat required to raise the temperature of water from room temperature to boiling point plus the heat required to transform water to vapour (steam) at 100°C (latent heat) is so great as to result in little loss of water during heating time. Indeed, a finely cooked muffin, irrespective of its composition, should still be moist after baking, a fact which implies retention of water in the food itself.

I have a coffee cup with a glass handle. I expect it to feel hot, but it is cool with the hottest coffee. Why is this?

It is a question of heat transfer, and how efficiently that occurs. Heat is lost in three ways, by conduction, convection and radiation. Conduction is relatively slow – heat moves through the material. Convection takes place when air is heated and carries heat away from a body. This is a much faster process when the air is flowing rapidly. Radiation is fastest but only important for really hot objects in comparison with others. For a warm or hot coffee cup, conduction is the most important means of losing heat.

It is a question of thermal conductivity. Some materials conduct heat away more effectively than others. No one would use an aluminium cup with an aluminium handle – you could not hold it!

A table of conductivities in relative units shows for:

Material	Conductivity	
Copper:	398	
Aluminium:	237	
Iron:	80	
Glass:	0.72	(500 times less than copper)
Wood (pine):	0.11	
Polystyrene foam:	0.033	

So a glass handle on a coffee cup is one of the best!

Why would two identical trays in a freezer be quite different in the ease with which ice cubes are removed from them?

Possibly because water from the tap or faucet on being run into a tray has absorbed different amounts of air in different cubes. The air has then been trapped in the frozen cubes. Ice is a heat insulator and air a conductor of heat, so a quick expansion of air might make the cubes jump from the tray on occasion. Also, even in a refrigerator, heat rises from the water as cooling occurs, causing a slightly different environment for upper and lower tray. Expansion of the ice as it freezes can force it out of its plastic container.

While dining in a restaurant in Toronto, a guest of mine began to open a bottle of ketchup which exploded sending a stream of sauce many metres away over other guests and an adjacent wall. How could this have happened?

It is not normal for a ketchup bottle to 'explode'. On the contrary, it is often difficult to persuade tomato ketchup to leave the bottle even after vigorous agitation or tapping has taken place. The situation basically is this: 'ketchup' is a puree that compresses somewhat when the bottle containing it is inverted. When this happens, a partial vacuum (absence of air) occurs at the bottom, and when the cap is removed, atmospheric pressure acting on the exposed ketchup surface supports the column of puree to the extent that it is usually difficult to initiate downward flow. Indeed, until an air space is created between the ketchup and the inside of the glass bottle, flow is unlikely to start at all. The purpose of tilting the bottle, inserting a knife between ketchup and glass, or twisting the bottle is to create a passage from outside the bottle to the evacuated space at the bottom (now the top) thus equalising pressures and making flow of the fluid more likely. On the other hand, the purpose behind hitting an inverted bottle violently on the bottom is merely to ensure by brute force that a downward force exerted manually, when added to the pull of gravity on the puree, is sufficient to overcome the role of atmospheric pressure in keeping the fluid fixed in the bottle.

An important general feature of a puree, like ketchup, is that it is a 'thixotropic' substance, one that is normally a solid of sorts but which when a shear force is applied to it will flow as a liquid. Ketchup, like margarine, can easily be spread with a knife (shear force) but becomes more solid once the spreading stops. Thixotropy means 'touching and turning' in Greek and so agitation can actually cause the liquid to flow when it otherwise would not do so. Ketchup is a sauce with principal ingredients tomato pulp, sugar and vinegar.

Bottles do, of course, sometimes 'explode'. One even did so on Canadian television several years ago during an interview. For this to occur pressures as much as ten times greater than atmospheric pressure are often required, and the bottling industry is very much aware of the dangers and hazards of aerated beverages when agitated, heated or treated in such a way that gas dissolved in liquid is removed from the liquid to form high pressure gas in the vacant space above it. A ketchup bottle, of course, is not a carbonated soft drink bottle. On the other hand, a build-up of gas in any bottle will create a positive pressure if it arises from a change from liquid or solid to vapour, or from a particular chemical reaction inside the bottle. What we learn from this fact is that the generation of a gas like carbon dioxide within a sealed container can have dramatic results.

How could this then occur in a ketchup bottle? The only possibility is fermentation. The end products of natural fermentation of sugars, namely carbon dioxide and alcohol, were identified first by Gay-Lussac in 1810, and the work of Pasteur in 1857 did much to explain the process. From the recipe for ketchup given above, it is clear that sugar figures prominently in its composition. It also has low acidity, and its preparation ensures that few micro-organisms or spores are trapped and survive inside the bottle. On the other hand, it seems that the presence of an appropriate mould or fungus in the bottle would be sufficient to generate gas from interaction with the sugar in the puree. Appropriate spores in the air surrounding an open bottle, particularly in a warm environment could make mould formation possible. There is also the possibility of bread crumbs (yeast) from a tablecloth falling into a used bottle which was then sealed, as an alternative to mould formation. Clearly the warmer the environment for storage of used ketchup, the exposure to air, and the practice in a restaurant of topping-up one bottle from another, could all contribute to the possibility of fermentation as the cause of the exploding ketchup problem.

Finally warm ketchup in an often-used bottle that is severely agitated, will spurt from it with alacrity if the top is not firmly in place. The bottle will not have 'exploded', but the results can be almost equally dramatic. A long distance of projection on the other hand would seem unlikely in this case.

Part Four
Miscellaneous

A radio listener uses a vacuum press to apply veneer to furniture. He places his sample of wood covered with resin and veneer in a vinyl bag and then applies pressure through evacuation of the bag. How does this work? What does the reading on the gauge mean (it says 30)?

It is clear that the questioner puts his piece of wood in a thin vinyl bag and then removes air from the bag with a vacuum pump. Now, with atmospheric pressure outside and a vacuum inside the bag collapses under pressure on to the work piece, bonding the veneer to the resin to the wood.

The gauge measures the extent to which pressure inside the bag is less than that outside. The number 30 with Hg after it indicates a pressure of 30 inches or 76 centimetres of mercury. Torricelli (1647) first used a mercury barometer to measure atmospheric pressure. 29.92 inches of mercury is equivalent to atmospheric pressure around 101.3 kilopascals. His gauge is a vacuum gauge and measures how much less than atmospheric pressure is the pressure in the bag. 30 would indicate 'no air' or a perfect vacuum, but no pump can actually produce that.

'Aliphatic' resins dry by moisture loss. The vacuum pump removes water and other vapours, leaving a vacuum dried and set resin.

Why does your singing sound more wonderful and possess more power in the shower?

Normally when you sing in an auditorium or outside you hear only the sound of your voice as you create it. In the shower the sounds you emit are repeated back many times from the walls or curtains and arrive at slightly different times. As a result the duration and overall intensity of sound is greater as you hear it. The continuation of sound gives brilliance to high notes and fullness to low ones. You become a star!

Is it true that when you take a shower large electrical fields can be set up in the air in the room?

It is true. Fields up to 800 volts per metre exist in the vicinity of the shower. It has been shown, though not understood, that large water droplets in the air round the shower are positively charged whereas small light aerosol-like drops are negatively charged. Gravity causes the large ones to fall, leaving the air negatively charged from small droplets. Similar negative ions are found near waterfalls. Cleaning petroleum trucks with a high-velocity water stream creates huge electrical fields and occasionally causes an accidental explosion.

Is it true that chloroform is to be found in domestic showers?

Yes. A recent medical conference has confirmed the existence of significant amounts of chloroform in showers. How it works is this – chlorine is used in many localities to purify water by killing micro-organisms. These little bugs being organic consist of carbon and hydrogen so that when killed by chlorine, the chlorine combines with carbon to form chloroform, chemically $HCCL_3$. The more micro-organisms in your unpurified water, the more chloroform in your shower. This, however, is not a great problem for most people. However, it is prudent not to breathe too deeply or sing too long in your shower. Deep breathing and endurance might increase the probability of succumbing to this toxicity from cleanliness.

Just how does a halogen quartz lamp differ from a normal source of light?

Incandescent lighting (light from a hot filament) has been around for most of this century. It comes from heating a wire to high temperature. Tungsten has always been the material of choice. However to get good illumination the tungsten has to be overheated. The material evaporates, thins, and then the wire burns through. End of lamp. New halogen quartz lamps also have a tungsten filament, but in this case it is enclosed in an atmosphere of iodine vapour. Evaporated tungsten however likes to combine with iodine to form a tungsten iodide. This vapour then, on coming in contact with the hottest and now the thinnest part of the original tungsten wire, breaks down and deposits tungsten back on the wire. This gives the lamp a long life and good output. This fact was first noted by Van Arkel (1920).

Because of the high temperature of the wire, a glass window would crack if of ordinary glass, while iodine would actually react with soda glass. Therefore a quartz window is used to transmit halogen lamp light, and effectively transmits ultraviolet as well as the visible light. These harmful UV rays can be absorbed by a plain soda glass envelope or window outside the lamp, and this is normally done.

What is the logic behind and the cost of 'degreening' oranges?

Chlorophyll is the green pigment that plants use to carry out the process of photosynthesis in which sunlight is absorbed and carbon dioxide changed into sugars and other plant materials. In much of the world the natural colour of a sweet orange is green. Only in cool weather does the skin turn orange and that colour has nothing to do with ripeness. Canadians however like their oranges to be orange-coloured and this fact is a costly luxury. Recently consumers have started asking themselves whether the colouring is worth the effort.

Usually the natural colour of a ripe orange is not orange. In the tropics all naturally ripened oranges are green. In subtropical regions however orange skin simply turns orange in cooler weather and the colour has nothing to do with the maturity of the fruit. Indeed 'orange' oranges often revert to a green colour as the temperature warms. The colour green merely indicates that photosynthesis is proceeding normally. In its absence the opposite is the case.

The oranges that we buy in supermarkets however are invariably orange in colour these days. The reason is simple. Because citrus fruit growers realise that consumers associate the colour green with immaturity in a fruit, they take over from nature and change the colour of sweet oranges. To do so they place the fruit in a de-greening room filled with ethylene gas which destroys chlorophyll. Alternatively oranges can be dipped and waxed to bring out the orange colour. In this case a non-toxic food colouring is used which destroys the chlorophyll. The coating of wax acts as a preservative and provides the shine consumers have come to expect of supermarket oranges.

The question of de-greening oranges arises periodically even today, largely because green oranges would cost less and have a longer than average shelf life as natural fruit. It seems unlikely however that they will take over the supermarket shelves for some time, because some twenty years ago the public was asked in a survey if they would buy Florida oranges which were ripe but green. Overwhelmingly they said yes. But, when green oranges from Florida came to the shelves they stayed there, whereas orange oranges from Israel, Spain and Brazil became highly popular although more expensive. Since then each year around ten million boxes of Florida oranges are sold in Canada – all de-greened.

How is it possible for ordinary humans to walk on hot coals, or is it?

It is possible, and even simple. Usually no great risk to life or limb is involved. The principle is straightforward and involves the vaporisation of water. Think of dropping drops of water on to a hot plate or skillet. The drops run around and gradually reduce in size. Indeed it takes up to two minutes for the water to evaporate. Because of this fact it takes a significant time for feet to heat and burn.

Our bodies are covered by a layer of water, on the surface of which is a layer of water-vapour. Hot coals will turn the film of water nearest them into vapour, and water vapour is a gas. The rest of the water is separated from the hot source by an insulating (to some extent) layer of gas. As a result heat finds it more difficult to get to the rest of the water layer. Conduction is poor, and convection is ineffective. As a result most heat is transferred by radiation, and this slows down the boiling of the layer of perspiration or sweat on the feet. The drops on the skillet vaporise at the bottom rapidly and the drops then become a hovercraft with a new layer of vapour inhibiting the flow of heat from below.

When something is red-hot its temperature is usually 300–500°C – more than enough to boil water.

So the time of contact is very important. Walking continuously at a regular rate ensures the ability to replenish the water layer between contacts. Spending more time in contact with coals increases the probability of harm.

Always be careful. The surface for walking should be firm and stable. Burning wood will set fire to leg hairs if the foot falls through the fire, and burning wood or hot ash will burn skin directly if placed on top of it. Make sure you have calculated all contact times carefully. Remember even the drops on the skillet only last for a minute or so.

Are 'phosphors' new materials?

No. The name is derived from phosphorous, an element common in organic matter – it is a non-metallic solid waxy substance which, as it oxidises in air, emits an eerie green light. Other substances which emit light, particularly after invisible light (ultraviolet) has fallen on them are called phosphors. Decaying trees often drip phosphorous through biochemical action. Phosphors are used as keyhole lights, absorbing sunlight by day and emitting light by night.

A medical specialist wants to know why, when he turns on the hot water tap, great banging and rumbling sounds ensue as the water turns from cold to hot.

The reason usually is that with increasing water flow through the restricted or narrow parts of a pipe system, turbulence occurs followed by cavitation – the creation of bubbles. For normal steady flow of a liquid, the speed of flow increases as the diameter of pipe decreases. For narrow streams, water speed is high and air is sucked into the channel. The noise heard is due to oscillations of air bubbles which emit sounds amplified by pipes, walls, ceilings and floors to which the pipes are connected.

Bubbles are continually being formed, expanding, oscillating and exploding. For low speeds of flow little air is sucked into the water stream. For others it depends upon the amount of air trapped in the narrow sections of pipe. The resonant noises you hear are similar to those from pouring out a water bottle, only magnified by the overall pipe system.

The problem can be alleviated by adding a vertical pipe of trapped air to the water pipe. The bubbles then leave the stream and the turbulence ceases.

What is physics?

Physics was mentioned by Thomas Young in 1807 in his classic Royal Institution Lecture Series: 'A Course of Lectures on Natural Philosophy and the Mechanical Arts'. He distinguished between 'physic' as a medical term and 'physics' as a branch of natural philosophy. His book started with mechanics and went on to other properties of matter while excluding all aspects of medicine and engineering which would previously have been included in such books. It was William Whewell in 1837 however who coined the word 'physicist'. In an essay he suggested two new words 'scientist' and 'physicist'. Michael Faraday, the eminent physicist and chemist (1791–1867), approved of the designation 'scientist' but not 'physicist'. 'Physicist is to my mouth and ears so awkward that I think I shall never be able to use it,' he said. Three separate 'i' sounds in one word is too much. However *Blackwood's Magazine* (1843) said 'its four sibilant consonants fizz like a firework'. Finally Tyndall, great scientist and the new President of the Royal Institution, actually referred to himself as a 'physicist' in his inaugural address, and life has never been the same since.

When I stand on a highly accurate weighing scale, I observe that the needle giving the result oscillates between a value that is higher and one that is lower than the normal average weight. Why would this be?

The heart is an organ that pumps blood containing the oxygen necessary for the continuing nourishment of the body. As the heart goes through its pumping cycle, the blood is moved up and down the body with the result that the centre of mass of the blood system is to be found at different locations as the cycle progresses.

The oscillation of the needle is small but can make a difference of 28 grams in a mass of 75 kilograms on either side of the normal average reading. This uncertainty is not a problem, especially as there is very little a person can do about it.

What is activated charcoal?

As its name suggests charcoal is a form of carbon obtained from burning or heating material containing carbon with limited access to air. The word 'char', just as in the barbecuing term 'char'broiled, suggests incomplete burning.

Charcoal burning is an ancient art. The charcoal burner's hut was a familiar sight in ancient times, and the heat generated enabled mixtures of materials to become glasses and other ornamental materials.

'Activated' charcoal is produced by special manufacturing techniques which allow the production of highly porous charcoals which are almost entirely 'surface'. Surface areas of 2,000 square metres per gram or 2 million per kilogram are possible. Surfaces adsorb or collect odours or coloured substances from gases and liquids. The large surface area allows the removal of huge amounts of gaseous material from the environment.

Activated charcoal is used for the purification of drinking water, in the recovery of paint solvents and other toxic liquids, and in gas masks for the removal of toxic compounds from air.

If you have an aquarium or fish tank, pellets of activated charcoal are regularly placed in the tank and remove impurities from the water. The charcoal first absorbs as much gas or odour as its surface will allow. Then, if it is to be reused it will first be taken to an appropriate place, heated and the gases released. The material can then be reused.

When a spacecraft is in an orbit around the earth and people and things within it are weightless, what happens to a glass of water in the craft?

If an open, clean, glass jar containing water is taken on board an orbiting spacecraft where it becomes weightless, the water climbs up the inner wall and down the outside. Water wets glass, and on earth in the case of water in a straw, the level of liquid rises until its weight equals the force exerted by molecular attraction between water and straw. When water is weightless, the molecular attraction between water and glass allows the water to climb up and over the jar.

If an astronaut in a spacecraft puts a kettle of water on an electric stove to boil under weightless conditions he finds the water on top of the kettle is still cool an hour later. How can that be?

On earth, water when heated at the bottom of the kettle becomes lighter (hotter and less dense), and is displaced by cold water which is heavier, sinks, gets heated, and continues the cycle. 'Convection' currents mix hot and cold water, and all the liquid is warmed. In weightless environments nothing is lighter or heavier than anything else. Water is heated by 'conduction', a very slow process involving only the next layer of liquid and proceeds layer by layer. Heating is a slow business.

How can an astronaut pour liquid from one container into another when weightless?

The easiest way is with plastic or elastic containers and squeezing liquid out like toothpaste. Another is to place two containers mouth-to-mouth and suddenly move them in a direction opposite to pouring. Conservation of momentum then solves the problem. The mass of liquid is greater than the containers, so its acceleration is less than the containers. It is as if the astronaut slides the first container off the liquid; the second on to it.

A questioner asks why she does not get an electric shock when she sprays water over a garment exhibiting static cling.

The question involves static electricity, a phenomenon well known in Manitoba but not in more humid areas. The science involved is the same as that involved in obtaining an electric shock after walking across a rug or sliding across a plastic car seat.

Ever since 600 BC when Greeks first picked up amber (a resin), rubbed it and found that it could pick up straw, grass or strips of metal, static electricity has been studied. Amber rubbed with a cloth or glass rubbed with cat's fur, exert attraction or repulsion on other objects. Rubbing removes electrons from one object and places them on another. All these materials are 'insulators' which do not conduct electricity, so charge (in dry conditions) will stay upon them. A 'conductor' on the other hand will conduct electric charge away as electricity.

Plastics are insulators and become charged due to rubbing. Spin-drying is likely to cause wrinkling for this reason. A spray of water, which is a poor conductor but not an insulator, will however allow charge to flow away, leaving the clothing wrinkle free. An aluminium film would do the same, only better.

I take an open bottle, fill it with water, and then insert a cork smaller in diameter than the mouth of the vessel. I hold it flush with the bottom of the bottle. The bottle, water and cork are then dropped from a great height under gravity and the cork is instantly released. What happens when the bottle falls freely? Does the cork move to the surface of the water?

No, it does not. Objects float when their weight equals the buoyant force upwards due to pressure differences in the liquid. In a bath a person sinks to the bottom as the weight is greater than the buoyant force, a denser liquid like the Dead Sea may have an upward buoyant force equal to the person's weight. A cork below ordinary water is being held there by an additional force which makes up for the difference between its weight and the buoyant force.

To simplify the question, assume you push the cork down to the bottom with your finger and hold it there until the bottle is dropped. Then it is as if you switched off gravity. Everything falls together – at least for a while – and there is no pressure difference between top and bottom of the liquid. Therefore no buoyant force exists to move the cork upwards. It, in fact, stays on the bottom. So what happened to the buoyant force you created by pushing the cork down?

Because water is slightly compressible, a small shock wave will dissipate energy instantly once the finger is removed, resulting in small oscillations of the cork. The cork, however, will not move upwards relative to the water.

Part Five

Fun Questions for Kids of All Ages

How can a physicist account for the exceptional redness of Rudolph, the Red-Nosed Reindeer's nose?

There seem to be two possible answers to this question: either the redness is in fact the radiation of a hot body that is being observed or, perhaps, the nose in question is actually red in colour. It should, however, be noted that in the university environment the two groups who claim to see Rudolph most readily are either naughty graduate students who stay up past midnight or professors still marking term papers on Christmas Eve, and these are among the least reliable groups in society as far as observation goes.

Both, however, claim to see red in *The Night Before Christmas* which indicates that the nose is actually radiating light; 'you could even say it glows'. An exhaustive analysis of this subject produces the fact that there is a structural similarity between the nose of a reindeer and the fibres of the nose-cone of a rocket. It may well be that in a similar manner to a rocket undergoing re-entry into the earth's atmosphere, the friction of the earth's atmosphere causes the reindeer's nose to radiate heat. Furthermore, if one makes a calculation involving friction, the specific heat of reindeer tissue, the thermal conductivity of hide, and the mass of a standard nose (taken as half a kilogram), it seems that an emission spectrum of red would be achieved at a velocity of something over 1,200 kilometres per hour. A reasonable conclusion would be that Rudolph is almost supersonic but not quite. After all, he has to stay below the speed of sound to hear Santa Claus's clear instructions. It can be assumed that only Rudolph has a red nose because only he can go that fast which, of course, would make him the harried Santa's favourite.

Why has Santa's hideout never been discovered by any expedition to the North Pole?

A puzzling fact contained in this question is that children to this day address Santa's letters, not to the North Pole, but to Greenland instead. A recent scholarly publication, in fact, has indicated that an American arctic station established in 1915 has, in the intervening period between then and today, drifted over two hundred miles toward the North Pole. It may be that continual drift of Santa's hideout would account both for the difficulty in detecting its whereabouts and its change of address. It would also account for the fact that Santa's workshop has never been observed by expeditions to the North Pole.

How can it be possible for Santa's team to deliver presents all around the world in one day and always by night?

The obvious answer to this question is that the reindeer space themselves at regular intervals along a meridian of the earth and maintain themselves in a stationary posture with respect to the fixed stars so that once in position, with the world revolving beneath them once every twenty-four hours, the task of the reindeer is simply to drop the presents at the right time, down the correct chimneys. The finding, several years ago, of numerous well-wrapped Christmas gifts in the middle of the Pacific Ocean adds some credence to this theory. Indeed, it shows that the deer are fallible and occasionally drop their delivery parcel in a region where there are no chimneys whatsoever. It is therefore the presence of fixed stars that makes the whole activity possible and it is clearly only the fallibility of the reindeer that results in certain chimneys being missed and other presents being located at unexpected destinations. It should also be noted, in relation to the Rudolph question, that his red nose is observed on Christmas Eve only because he is leading the team to their fixed positions and reaches a speed that is almost supersonic as he takes his position.

A Merry Christmas to all our readers!

Glossary

albumen	The white of an egg consisting mostly of albumin dissolved in water.
albumin	Any of a class of proteins that are soluble in water and can be coagulated by heat, found in the white of an egg.
aliphatic	Belonging to a class of organic compounds in which the carbon atoms form chains with open ends rather than rings.
buoyancy	A body's loss in weight or tendency to float when immersed in a fluid (buoyant force).
cavitation	The formation of cavities in a fluid downstream from an object moving in it, as behind the moving blades of a propeller (bubbles).
centre of mass	That point in a body which moves as though it bore the entire mass of the body, usually identical with the centre of gravity.
ceramic	A product made by the baking or firing of a non-metallic mineral, such as tile, cement, plastic refractory, or brick.
charcoal	A porous solid product containing 85-90% carbon and produced by heating carbonaceous materials such as cellulose, wood, peat, potatoes or peel at 500-600°C in the absence of air.

chlorophyll	The colouring matter of the leaves and other green parts of plants occurring in small bodies within a cell.
colloid	A substance composed of particles that are extremely small but larger than most molecules. The particles in the colloid do not actually dissolve but remain suspended in a suitable gas, liquid or solid.
convection	The transfer of heat from one place to another by the circulation of currents of heated particles of a gas or liquid due to a temperature gradient.
dew	The fine water droplets, condensed on solid surfaces when the air temperature drops below the dew point.
dew point	The temperature to which air, at a given pressure and water vapour content, must be lowered for saturation to occur, i.e. the temperature at which dew forms.
electrochemical	Having to do with that branch of chemistry that deals with chemical changes produced by electricity and the production of electricity by chemical change.
equinox	Either of the two times in the year when the centre of the sun crosses the celestial equator, and day and night are of equal length in all parts of the earth; occurring in the northern hemisphere about 21st March (vernal equinox), and about 23rd September (autumnal equinox).
freeze-drying	The dehydration or removal of water

	by freezing the moisture content to ice and evaporating the ice in a vacuum.
galvanic cell	An electrolytic cell for the production of electricity by chemical action.
gel	A jelly-like or solid material formed from a colloidal solution. When glue sets, it forms a gel. Egg white gels when it is cooked.
halogen	Any of the group of elements that consists of fluorine, chlorine, bromine, iodine, astatine.
heat capacity	The ratio of the quantity of heat applied to a body to the change in temperature the heat induces, usually expressed in joules per degree Celsius (alternatively, calories per degree Celsius).
Hertz (Hz)	The unit of frequency, equivalent to one cycle per second.
humidity	The amount of moisture or water vapour in the atmosphere.
relative humidity	The ratio between the amount of water vapour present in the air and the maximum the air could hold at the same temperature.
infra-red	The invisible part of the spectrum whose rays have wavelengths longer than those of the red and the visible spectrum and shorter than those of microwaves.
insulator	Something that inhibits the passage of electricity, heat, or sound.
lactic acid	A colourless, odourless acid produced by muscle tissue during exercise.

latent heat	The heat required to change a solid to a liquid or a vapour, or to change a liquid to a vapour without a change of temperature.
lime	A solid, white compound of calcium and oxygen obtained by heating limestone, shells, bones, or other forms of calcium carbonate.
meridian	A half-circle extending from one pole to the other on the earth's surface is known as a meridian.
microwave	A high-frequency electromagnetic wave, usually one millimetre to one metre in wavelength. A microwave oven operates at 2450 MHz giving a wavelength of 12 cm to the radiation.
octagonal	Having eight sides.
phosphorescence	The emission of light without noticeable heat and without burning.
photosynthesis	The process by which plant cells make carbohydrates by combining carbon dioxide and water in the presence of chlorophyll and light, and release oxygen as a by-product.
Polaroid	A commonly used trade name for material that will cause polarisation. Polarising materials transmit components of light that lie in one vibration direction and hold back all the others by absorbing them internally.
polystyrene	A water-white, tough synthetic resin made by polymerisation of styrene. It can be manufactured as a clear film.
quartz glass	A clear, vitreous solid formed from pure quartz that has been melted. It is characterised by an ability to with-

stand large and quick temperature changes, is chemically inert, has a high melting point, and is specially transparent to visible and ultraviolet radiation.

radiation — The act or process of emitting energy in the form of rays or waves, especially electromagnetic waves.

saturated — Something that has combined with or taken up in solution, the largest possible portion of some other substance.

shear — Force causing two parts or pieces to slide on each other in opposite directions.

soda ash — Sodium carbonate in a powdery white form, partly purified for commercial use.

solstice — Either of the two times of the year when the sun is at its greatest distance from the celestial equator. In the northern hemisphere 21st or 22nd June, the summer solstice is the longest day of the year and 21st or 22nd December, the winter solstice, the shortest.

specific heat — The number of joules of heat energy required to raise the temperature of one kilogram of a substance one degree Celsius (alternatively, the number of calories of heat required to raise the temperature of one gram of a substance one degree Celsius).

stress — The internal resistance to an external force such as tension or shear which tends to cause a change in the shape or

	volume of a body, expressed as the force per unit area acting on the body.
sundial	An instrument for telling time by the sun; it is composed of a stylus that casts a shadow and a dial on which hour lines are marked and upon which the shadow falls.
temperature inversion	A layer in the atmosphere in which temperature increases with altitude, the principal characteristic of an inversion layer is its marked static stability, so that very little turbulent exchange can occur within it.
thermal conductivity	The heat flow across a surface per unit area per unit time, divided by the negative of the rate of change in temperature with distance in a direction perpendicular to the surface, also known as a coefficient of conductivity or heat conductivity.
thixotropy	The property of certain gels which liquefy when subjected to vibratory forces or even a simple shake, and then solidify again when left standing.
turbulence	The motion of fluids in which local velocities and pressures fluctuate regularly, in a random manner causing mixing.
ultraviolet	Having to do with the invisible part of the spectrum whose rays have wavelengths shorter than those of the violet end of the visible spectrum and longer than those of the x-rays. Ultraviolet wavelengths extend from about 50-380 nanometres.
vacuum	A space completely devoid of matter.

An enclosed space from which all air or other gas has been removed.

vinyl

Vinyl plastic is any resin or polymer derived from vinyl monomers. Vinyl acetate resin is a colourless, odourless, light, stable, thermal plastic that is unaffected by water, gasoline, or oil.

visco-elasticity

A condition of having the properties of viscosity and elasticity together. Silicone putty is visco-elastic. A ball of it will bounce, but when left on the table for a few hours the same ball will flow under the force of gravity into a pancake shape.

wavelength

The distance between one peak or crest of a wave of light, heat, or other electromagnetic energy and the next corresponding peak or crest; the distance between particles that are in the same phase at the same time, measured in the direction in which the wave was travelling.